Lena Stromberg

Resting in celestial orbits

a book about dumb-bells, Aurora-signals and REM

LAP LAMBERT Academic Publishing

Imprint

Any brand names and product names mentioned in this book are subject to trademark, brand or patent protection and are trademarks or registered trademarks of their respective holders. The use of brand names, product names, common names, trade names, product descriptions etc. even without a particular marking in this work is in no way to be construed to mean that such names may be regarded as unrestricted in respect of trademark and brand protection legislation and could thus be used by anyone.

Cover image: www.ingimage.com

Publisher:
LAP LAMBERT Academic Publishing
is a trademark of
International Book Market Service Ltd., member of OmniScriptum Publishing Group
17 Meldrum Street, Beau Bassin 71504, Mauritius

Printed at: see last page
ISBN: 978-620-0-53292-3

Resting in celestial orbits

a book about dumb-bells, Aurora-signals and REM

Lena J-T Strömberg

PREFACE

Beside the subjects in the headings, the book contains related issues in terms of flows, biodynamics and atmospheres. These are described in Chapters accomplished by Figure legends.

The first chapter comprises boundary layers e.g. intersections of body parts. The word 'aura' is used when referring to aspects and properties related to the human (body). A state law in thermodynamics is analysed and phases e.g. snow & other materialisations of orbits are illustrated.
Knowledge about the atmosphere is important in our time when details about the climate has a great value.

In Chapter 2, isomorphisms (cohomologies) of a matrix state are proposed.

In the third chapter, orientations of dumb-bells for the body and sub-bodies are recognized and exemplified with e.g. sleep. Most people fall asleep not supine, but when lying with the head on one side. This changes the orientations of the eyes into vertical alignment.

Another action, observed on an acquaintance is to fold the head (and body) in a blanket. That may change chemical properties and pressure, such that a new state dominates. In other words, a canonical set of variables within thermodynamics rules, and gravity is not so present. Therefore it could be rest, although it does not look neither healthy, nor safe.

Also, the respiration when lying down gets a different direction and that may indicate a change for the eyes as a signal to fall asleep.

The senses are 5, but often a 6th is addressed, for example gut feeling. That is an ability to diagnose a situation and act accordingly. If is more related to surroundings than the own body. A related issue is balance which incorporate several senses. It is integrative physiology and succeeding may be supported by an acoustic activity: In a book of Sophie Kinsella, it is told that walking on high heels works when singing ' Land of hope and glory'.

In Chapter 4, it is emphasized (and established), that dynamical system may interact with a surrounding into organisation and memory. With those preliminaries, some dumb-bells for collection and redistribution of energy are gathered.

In addition, the couples dancing on the book cover could be considered dumb-bells in interaction. In view of this, one may imagine other civilizations where motions in a vertical alignment are more easily accomplished.

Whilst the first 4 Chapters contain amounts of mindfulness, the last treats more classical applications in solid mechanics.

Table of contents

1.

Flows in the eccentricity zone of a noncircular orbit, related to streamlines, boundary layers, temperature & molecule weight

Lena J-T Strömberg
previously Dep of Solid Mechanics, Royal Institute of Technology, KTH
lena_str@hotmail.com

Abstract. A theoretical description with exchange of energy in boundary layers is outlined for noncircular orbits. Stream line solutions with pressure, temperature and gravity are analysed. Both inertia energy and potential energy are considered in redistributions, mainly assuming an additive decomposition in a primary motion.

Keywords: Aurora, signal, climate, boundary layer, aura, configurational force, Hamiltonian, Euler-Lagrange, material force, pressure, temperature, molecule weight, noncircular orbit, nco, eccentricity zone, potential, gravity, centripetal acceleration, physiology, heat radiation, race track, Goldstein, Jabberwoky

1. Introduction

In an article from 2015, we raised the question about interactions with the Sun. Since it is so distant, it is likely that it pastes and delivers sunshine in an objective fashion. However there may be interactions, e.g. same as our tidal effects, and we can spot the eyes of the surface of the Sun, but these are most likely only for local interaction. Another possibility is that the atmosphere may copy the increasing dynamic activity of illumination and waves creating a larger scale   of   a   fractal.

The Sun beneath a cloud, gives an immediate response on Sun Breeze waves in   water.
A scenario, that may be present at Earth, is an escalating Sun like behaviour due to a combination of increased temperature and curls by electromagnetism on different levels.

From this, there is a possibility that Earth and Sun arrange into vertical dumb bell interaction, to obtain the same properties. Nor this is so likely, since the

Earth is diverted into several substantive atmospheric systems with their own frequencies and independency.

With Sun as a primary, the dumb bell may consist of Earth and a closer celestial object. Before Earth achieved its present orbital period, it is likely that it was a dumb bell with Venus, since Venus rotates in the opposite direction. The recent Venus passage may have recalled a material memory of that time, resulting in a higher temperature.

In an analogy with a tide, it may collect something related to Sun Shine. It could be interior, as the magma, known from volcano eruptions. There could be something in the atmosphere (compare so-called solar storms), that releases with periodicities of several years, so that ones in a while, there are hot summers.

In agreement with the results of some climate research, the weather localises into more extreme, i.e. colder, warmer and more rain and snow, when   those   events   occur.
Here, parts of that will be confirmed and described in more details.

In general, it is of interest to know how Earth and climate respond to extreme temperatures. Another issue is how temperature is measured and if we have absolute objective equipment. Predictions from modelling may contribute to knowledge and understanding. In the present context, concepts of heat and flow in continuum mechanics will be analysed for motions in *noncircular orbits*, nco.

To describe material behaviour in boundary value problems, BVP, there are models with homogenisation of properties on a minor scale. Hereby, known processes such as diffusion and slip line orientation are smeared to obtain mean values or material modules which are defined on a larger scale. Another, somewhat related, discipline is to introduce boundary layers. It is found that the position where the action occurs, often changes shape and develops instead of only transmit the forces and displacements as predicted in first order theory. That is the case at chaos, and it may be accomplished by structures which introduce growth in the layers, c.f. Appendix. With a collecting name, the phenomenon is called fractal.

Also In classical mathematics, the boundary conditions for higher order problems may be certain differential equation, e.g. the Robin condition, which is a sum of Dirichlet and Neumann.

The main results concern boundary locations and redistribution of forces and potentials to a surrounding so-called *eccentricity zone*. Such a rearrangement agrees with the concept of configurational forces in Euler Lagrange equations.

2. Distribution of forces into an 'aura'

With Hamiltonian mechanics as the point of departure, an additive decomposition of forces with potential will be introduced. Then, tacitly assuming a boundary layer, parts of the potential are transferred to other adjacent systems. These may be material bodies or a region with density. That surrounding is denoted as the *aura* of the primary system. With these preliminaries, the framework will be applied to (differentiation of) Bernoulli's law, to obtain physics and possible characteristic features of a noncircular orbit, given in Figure 1.

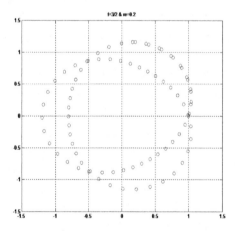

Figure 1. Radius vector r =r_0+r_e sin(fωt) for a noncircular orbit, with f=3/2, r_0 =1 and r_e =0.2. The surrounding eccentricity zone defined by r- r_0 will occasionally be denoted an aura.

Remarks. The aura-concept is more often used not so exact, whereas in science, when measures and quantities can be defined, a spatial boundary layer is introduced.

Reflexion of light in a surface, c.f. Figure 2, is due physics in an extended layer.

Figure 2. Lights reflecting in the glass-surfaces.

In a design for fabrics, an aura is pictured as curves, Figure 3.

Figure 3. Aura, imaged as noncircular orbits in a design at Spoonflower.

3. Kinematics of planetary rotations and free fall

First, we shall characterise known motions of rotations adjacent to a primary, and relate to an aura.

Definition: A synchronised rotation, is when the moon or planet show same side to the planet all time.

Proposition 3.1:
A co-rotating body, with the same side towards the primary, is in a synchronised rotation and the ratio orbital to sidereal is 1:1. Hereby, in such a rotation, the absolute rotation is nonzero.

Proof. Readily obtained from rigid body mechanics.

Corollary. Therefore, at free fall on Earth, there is also a rotational motion.

Remark. It is possible that this is subject to small oscillations, and the details in terms of forces and equation of motion is in agreement with the motions in Avd [1] and the aura concept.

Proposition 3.2: If the planet/moon rolls in its orbital motion, then this motion contributes to the sidereal in a way dependent on radius (and is determined based upon assumptions in each case).

4. Bernoulli's law for motions in arcs/trajectories of noncircular orbits

Bernoulli's law, BL, reads
$$p+\rho u^2/2+V_{pot}=\text{const} \qquad (1)$$
where p is pressure, ρ is the density, u is the velocity of the stream line and V_{pot} are non-specified additional potentials.

For example with p being the pressure at depth h, or the gravity potential, BL resembles a specific law for energy balance and conservation of energy. However, since the derivation concerns a stream line and gravity is often present in V_{pot}, the pressure could be interpreted as an additional specific potential energy. It could be external due to e.g. mechanics at boundaries to the material point or internal due to e.g. thermal properties. Primarily, the latter will be assumed here, such that p obeys the state law of an ideal gas
$$p=r\rho\theta \text{ where } r=k_B/m_w \qquad (2)$$
with k_B being Boltzmann's constant, m_w molecule-weight and θ temperature. Concerning the parameter r, a discussion about specific pressure laws was initiated at Researchgate by Dr Jerry Decker, and the correspondence above, between parameters and Reynolds number, was derived.

Remark. In a more detailed approach, one could derive candidates for governing differential equations for motions, with configurational forces at various assumptions for spatial dependency of pressure. The pressure term could possibly be decomposed into two terms. These will be an internal part and configurational force, where the latter is tacitly assumed to emanate from boundary layer interaction.

Preliminaries. The spatial stream-line is assumed to be smeared into an *eccentricity zone* of a noncircular orbit. Hereby, the velocity is generalised and in a first approximation it consists of a rigid body rotational velocity and a varying part. In general, we assume chaotic solutions, such that frequencies and oscillation can rule. This could be described with either an oscillating radius or an angular velocity $\omega(t)$ as in [1]

$$\omega(t) = \omega_0 \exp(-2(r_e/r_0)\sin(f\omega_0 t)) \qquad (3)$$

4.1. Analysis and discussion of the state law (2)

Heat in state law
In a project on Researchgate [3], Dr Jerry Decker characterizes heat: 'Heat can be conductive, convective, or radiant. When radiant it has a wave length longer then visible light. Heat is measured by calorimetry, temperature change in a standard material.'

First, we will scrutinise the state law as it appears in continuum mechanics c.f. (2) above.
In conjunction with BL, it gives that temperature decreases while a velocity, which agrees with e.g. windy weather conditions. By itself, i.e. at constant velocity, the term remains constant for an increased temperature at increased molecule weight. This agrees with how we experience temperature at larger humidity. Then, a larger molecule weight can be a model of air with water content instead of only air.

Heat perception.
The manner in how we detect heat and temperature with our own bodies, is related to skin physiology and internal physiology, mainly blood vessels, the vascular system and respiratory. For example dogs have the ability to control

their temperature by hyperventilation. The effect is probably two-folded, such that

1) the blood temperature is decreased at the tongue, where a convective flow is   present

2) there is a redistribution of body heat energy into another kinetic format of motions by the lungs through respiration.

Additional effects of heat are drying, and from what is known for humans, a late sign (of loss of body fat which contain water) is the increased concave form at the eyes, where the skeleton has holes. Possibly, this is a way of the body to handle radiant heat waves in a fashion suitable and useful for the individual.

With the energy of BL (1) considered as a Hamiltonian, it is seen that for an increased velocity, the temperature decreases, however it may also be a transfer to a larger Hamiltonian i.e. a higher total energy. This, and certain redistributions of potentials to the surroundings, will be described in the next section.

5. Hamiltonian principle and eccentricity zone

The aura region discussed above could be regarded as a boundary layer. Next, we will propose that the aura conditions, such as the forces acting on an adjacent body/object, can be derived from a Hamiltonian principle. With a semi-inverse method, the inertia part may be expressed in spatial variables (or other general suitable coordinates). In the Hamiltonian, parts of this could be included and other part may be shared to a surrounding. Hereby it will be present in the Euler-Lagrange equations as an exterior force. It could be included in the Lagrangian since the format has a potential, but the idea here is to identify such forces and interpret the format into physics.

5.1. Distribution of centripetal acceleration energy to a surrounding aura

The sign determines if the direction is inward or outwards. If it corresponded to a force, the outward would be a centrifugal force and the inward a gravitational. On a planet, a larger outward force may cause an escape from the bounded orbit, which is also related to hyperbolic solutions. Next, a kinematic statement will be proposed.

Proposal 5.1. If the aura affects the sidereal rotation, then an initial escape path may rearrange into sidereal rotation in the opposite direction.

Remarks. This is in agreement with geometry of rotations, e.g. Venus.

The description with angular velocity alone as in (3), describes the behaviour if including a direction of w. Another description is to consider the time varying part of (3), for discrete times, when negative.

5.2 Additional about dependency and molecule weight m_w

Lower temperature, admits higher velocity. Higher molecule weight has the same influence, which could be an effect similar to gravity if the aura assumptions are considered for a rotating stream. If an increased molecule weight is spread as an aura, it gives the kinematics of a rotating whirl which increase its diameter or feeds smaller whirls at its boundaries. For a planetary system, the sidereal rotation of satellites may correspond to the smaller whirls.

6. Electromagnetism and (radio-)signals

Magnetism appears to 'perform' better at low temperatures and in some cases, e.g. super fluidity, it conquers gravity. Hereby, it is possible that the rotations in conjunction with electromagnetic atmospheric waves, make the temperature additionally lower (compared with that predicted in classical BL alone) by a redistribution of heat to surroundings. This, since the phenomenon may be quantized, such that it needs a certain discrete exchange of matter or heat. That will make the environmental temperature higher, while hot. When cold, possibly the electromagnetism could work, but if there is no matter in the air to attach on, (particles in motion, compare e.g. gas dynamics), then it takes heat or matter (e.g. humidity) from a surrounding. Possibly, it may also promote the creation of snow, since snowflakes have shapes which agrees with materialisation of electromagnetism.

In Figure A2, hexagon snowflakes are shown. A discussion of solutions to such shapes is outlined in Appendix.

These explanations agree with climate changes, however radio waves and electromagnetism have been present for a long time. On the other hand, it now covers a great part of Earth. The transportation or reflexion of signals is well known, and a reference from 1993 tells about several routes. A not so common, since the noise part was too dominating, yet longrange; 1000km, was the Northern light; *Aurora*.

Another feature, that may rule the temperature around the planet, is a celestial mean, determined by the location in the solar system.

From knowledge of acoustic and electromagnetic waves, they land and fit on curved shapes, and could be induced in a fractal manner to atmospheric levels. For Earth, since subjected to drying heat and drying wave motions, both the flat areas, the curved river and lakes with low water levels, could lose its water. The crucial thing is the speed of this, and nowadays we know the exponential growth and polarisation into extreme.

If electromagnetism feeds on humidity, as suggested above, it is a question what it leaves as a result, and that is of course measurable and readily visual to be observed. Depending on how large a surrounding in space-time, it could be characterised as information to humans and universe, which again results in some water.

As for other dangers in history, e.g. nitro-glycerine, Mother Nature has found a solution to stabilize. New engineering constructions to withstand the catastrophes may be developed.

The most crucial scenario is a vertical dumb bell to Sun, as was discussed above in the introduction. Finally, we will devote a chapter on how to prevent that.

7. Analysis of climate changes and celestial actions to stabilize

An action would be to reduce electromagnetism. That will be difficult, because so much used worldwide since it is free and provided by communities.

Other actions would be to pinpoint, magnify and develop, the celestial properties of Earth which distinguish it from a Sun. These may be: 1) Tidal effects, which possibly is related to the presence of a moon, but ruled by twice the sidereal rotation. Perhaps, this could be magnified in a beneficial manner.

2) To address the closer, colder planet Mars. This could be by launching a large satellite with 3 times supersyncronized speed, similar to the inner moon Phobos. Equally large as Phobos will be technically impossible, but the idea should work also small scale, since the vertical dumb-bell is not (so very) size dependent. Also many small two-moon configurations as Phobos-Demios could work, since that may induce the larger scale real satellite orbits in an upper scale fractal (provided such celestial creations are possible). Then, instead of behaving more like a Sun, the Earth will turn to Mars-behaviour in terms of overall celestial manners, i.e. colder, more decoupled and individual as a usual Earth planet as we know it.

8. Other observations and issues for future work

An observation, is that chemical substances in nature may have developed, in kinds of eccentricity zones, by growth and repeating itself under sufficient pressure or temperature variations. Newton, as well as many other scientists occasionally turned their efforts to alchemy, trying to understand and copy the way gold is created in nature.

The particle wave duality does not show in the concrete manner as derived in [4]. The oscillations in a nco, may be rewritten into a wave that surrounds a rotating particle. Additional specifications are needed since the aura consists of two eccentricity zones, namely the orbital of the primary and the sidereal of the particle planet. A glance of the outcome of an analysis shows that this leaves lots of possibilities, such that measurements and other disturbances might dominate making an objective detection impossible.

- Consider Tti as a Hamiltonian, and derive Euler-Lagrange format with a semi-inverse method

- Identify BL in 3 dimensions as discussed in [2] and consider a projection onto planes

- Express ω^2 in density and the universal constant of gravity G, and compare with BL

- Formulate with nondimensional variables, and concern examples with numerical values to show levels

- It would be interesting to test race tracks with the shape of nco, e.g. with f=3 as seen in Figure 4. Specifically, that could be fast speed running for humans, horses or whippets, and strengthening effects due to natural oscillations are likely to be diagnosed.

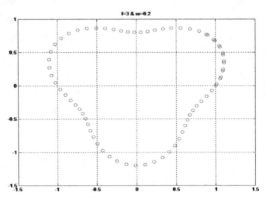

Figure 4. Noncircular orbit, nco with f=3. (Compare Figure 1)

9. Concluding summary

With planetary motion and whirly flow as a point of departure, we argued for a theoretical description with exchange of energy in boundary layers. These are assumed as spatially distributed in segments adjacent to rotating bodies. Both inertia energy and potential energy were considered in redistributions, mainly assuming an additive decomposition in a primary motion.

The energies were formulated with density, and for the case when Bernoulli's law (1) is satisfied on a stream line. Then, since densities appear on all terms, these are not present in the final expression. However in the state law for an ideal gas, continuum version, the molecule weight appears, c.f. equation (2). Hereby, as shown (with differentiation) above in subsection 4.1, a change in weight by e.g. densification has influence on behaviour of rotational velocity energy and specific gravity force-energy.

Acknowledgements

The author is most grateful to Adel Somalia for discussions on electromagnetism, light and duality, c.f. Section 8. Thanks are also due to Dr Jerry Decker, addressing chemistry in continuum mechanics, c.f. Section 4, and quoted in 4.1.

Appendix

Rotational motion of (almost) rigid bodies in interaction with boundary layers

Consider the rotation of a box, Illustrated in Figure A1. In a first approximation, the motion of center of mass and rotation are decoupled. With the side lengths a,b,c and a>b>c, the box has 3 nonequal inertia moments. Therefore, rotational motion around the mid axis (denoted 2) is chaotic.

In many books of mechanics, they consider the problem solved with the inertia ellipsoidal. That might be true for very or 'sufficiently' small times, but leaves questions, e.g. the loss of conservation of angular moment of momentum and how the angular velocity varies in time while on pole 'circles ''.

If the angular velocity is composed of tangent vectors, it is no longer continous, and then time has a nonzero minimum width, i.e. space between points. In real world, the box appears to move in continous rotation, occasionally, and that is the basis of the description leading to the Poinsot's ellipsoidal and pole hood. The case when two inertia moments equals is a precession invoking a harmonic oscillation around a spatial axis.

In a book by Goldstein [5], Poinsot's construction is accomplished by a footnote, 'notifying' it as a Jabberwokkan sounding statement. This might refer to Alice in Wonderland Figure A1 right, featuring a ghastly large wizard-monster Jabberwoky.

Figure A1. Left. Rotational motion of a box around its axis with mid inertia moment.   Right. Illustrations of Alice in wonderland without Jabberwoky

A snowflake, at growth is not symmetric before it gets all the corners, c.f. Figure A2.

That probably induces and promotes the fractal growth such that boundary layers of humid air become the characteristic structure of a snowflake. Also other more organised crystal growth are possible at ideal conditions, http://www.snowcrystals.com/videos/videos.html

Among several other paintings and scetches, da Vinci draw a human in a circle, somewhat indicating a pentagon shape, Figure A2.

Figure A2. Left. Snowflakes of various kind, pentagon shape and fractal. Right. Art by Leonardo da Vinci including and emphasizing science.

2.

The Matlab- and Octave-matrix magic(3) applied to nonlinear dynamics and stability

Abstract. Applications of magic(3) related to nonlinear dynamics are proposed.

Keywords: magic(3), constraint, Matlab, Octave-online, eigen-values, pressure, stability, buckling, structures, materialisation, tree, udder, states, bifurcations

1. Introduction

Concepts of magic(n), found in [1] are elaborated. Short summary and preliminaries: The matrix A=magic(n) in Matlab and Octave-online is nxn and characterised by that the first invariant tr(A) and the sum of elements in each row and column, equal the same constant. These sums will be known as the constraints.

In the present context, we address magic(3) in 3 different decompositions, namely the original, the diagonalised with eigenvalues and an additive decomposition with a hydrostatic state and remainder.

The word cohomology is used meaning when a shape in space(time) is in analogy with a flow, a stationary state or a physical space described with mathematics. Here, also the word materialize will be used.

Since the trace of the stress tensor is a pressure and pressure is balanced with energy in Bernoulli's law, the other sums may be considered as composed of various energies e.g. kinetic, potential, (heat).

The constraints in magic(n) is a reduction of d.o.f. That idea is sometimes exploited in theory of chaos, as a rule when there are many bifurcation providing several other options. For that reason, we will discuss magic(n) as a refined model for the scenario at loss of stability. Two examples are outlined, namely

- initially in-plane William's toggle frame
- buckling of a soda can

Then, the features of magic(3) will be analysed in terms of materialisation and growth. In modern mathematics, the discipline topology bridges certain mathematical properties with spatial properties of bodies, e.g. the torus, c.f. [2],[3]. Although not so concrete, in terms of functional relations, the methods are readily used and plausible.

The states will be found to materialize as a tree. As another example where the different decompositions occur, we will consider an udder.

Finally, and not so serious, a more chaotic configuration at Times Square is outlined.

2. Matlab-calculations

A visualization of A and a symmetric part is given by meshing the value at its position in a Cartesian coordinate system, c.f. the plots below copied from [1].

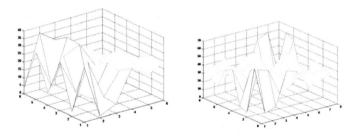

Left. mesh(magic(6));. Right. Mesh of symmetrized matrix for n=8; (magic(8)+ magic(8)')/2

Following results are found in Appendix A: For n=3, and neglecting the non-symmetric part, there is a coordinate orientation such that A= diag (15,-5,5)

This may be decomposed into a hydrostatic state and a '+- ', that reads

$$A= 5\text{eye}(3) + 10e_1e_1^T - 10e_2e_2^T \qquad (1)$$

where eye(3) denotes the unity matrix.

The non-symmetric part, in some examples below regarded as (a rigid body) rotation, is given in Appendix A, and provides rotations around all three axes.

3. Stability. Snap through in a beam frame and buckling of a soda can

A curved beam with an applied force or deflection P,v at the center is known as a Williams toggle frame, c.f. Figure 1.

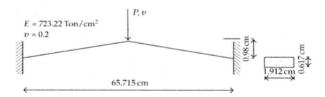

Figure 1. William's toggle frame: Inplane beam with applied deflection v.

Since initially bent, the force in the center provides compression on the upper side and tension on the lower side. In axial direction, i.e. horisontal in the figure, there will be an axial compression, corresponding to the largest eigenvalue in magic(n). The three values in orthogonal directions provides the symmetric part of magic(3). At snap through, the entire frame rotates out of plane and the center point rotates around a perpendicular axis, if the second mode is reached. This, together with the axial rotation and rotation at hinges is the skew part of magic(n). Hereby, an extra dimension is present at the snap through.

In three dimensions, buckling of a soda can, subjected to preload from axial compression, Figure 2, a similar scenario may occur.

As another scenario, we will assume an interior vacuum state inside the thin membrane. Then, an applied force at each side may correspond to the so-called cosmic ray, assumed to initiate thunder. The can is too constrained to achieve rigid body rotations itself, but at collapse, there are rotations at smaller scales.

Figure 2. Buckling of a soda can. The impact forces on each side are not shown in the figure.

Comparing with the values and directions derived in Section2, we may note that the impact force increases from 10 to 15, if the state of magic(3) changes from pressure to the diagonalised. This promotes the fast buckling collapse.

4. Materialisation into a tree and an udder

Consider an initially bend upwards of the tree branches. Then, additional bending upwards and downwards gives the + - eigenvalues, respectively. This, together with a larger tension inside the tree bole, gives the states consistent with the symmetric part of magic(3) in its diagonal form.   E.g. the leafs, subjected to rotations may represent the skew part.

All outer surfaces subjected to only a pressure, together with the above bending gives a cohomology with magic(3) in the state (1) as defined in Section 2 above. As a shape, the hydrostatic state may materialise as a spherical crown at the top of a tree, and by on large the root system may be considered a sphere. A transition to higher dimensions, could be thought of as a fractal contact with water pressure on several exposed surfaces. In Figure 3, parts of an old beautiful tree are shown.

Figure 3. A 200 year old Wisteria tree in Japan.

A cohomology with an udder, Figure 4 is even more direct: The inside milk provides a hydrostatic pressure. Then, the action on the teats/dugs provides a compression and a deformation-induced tension. When the size matches the pressure value of magic(3) in state (1), the state may rearrange from pressure into a stream, at which a lower compression/tension is required. The skew part may correspond to rotations of the inside milk.

Figure 4. Illustrations of an udder.

This observation provides a key to understand the chaos principle into minimisation of dof: When several bifurcations, a system may create new dynamics by changing between states, in order to interact with surroundings when suitable and in consistency with energy balance.

6. Conclusion

A reduction of d.o.f is sometimes assumed as a governing principle at chaos. In nonlinear systems, the unstable state could be considered chaotic when there are more than one solution and a more detailed description would lead to higher degrees of non-linearity and so on, although sometimes the requirements for the notation 'entire chaos' are that of several bifurcations. When there are changes between states at a finer scale, the system creates dynamics and may interact with surroundings by e.g. frequency coupling and other energy exchange.

Acknowledgements

To Anders, Maria and new aquaintanceship on social media.

Appendix A

```
octave:1> magic(3)
ans =
   8   1   6
   3   5   7
   4   9   2
```

Skew part of magic(3) reads

```
octave:2> (magic(3)-magic(3)')/2
ans =
   0  -1   1
   1   0  -1
  -1   1   0
octave:3> (magic(3)+magic(3)')/2
ans =
   8   2   5
   2   5   8
   5   8   2
octave:7> eig((magic(3)+magic(3)')/2)
ans =
  -5.1962
   5.1962
  15.0000
octave:8> eig(magic(3))
ans =
  15.0000
   4.8990
  -4.8990
```

3.

Resting in reversed dumb bell orientations

Summary. The symmetry breaking at sleep is analysed in terms of eigen-frequencies and resonance for dumb bells. Possible image processing at REM and reversed orientations are discussed. Reversed symmetry actions are proposed for resting of horisontal dumb bells.

Keywords. eigen frequency, oscillation, dumb bell, dectivated dumb bell, fold hands, rest, sleep, recovery, cycle, REM, vertical alignment, resonance, 'related resonance', reconfiguration, static positions

1. Introduction

Dumb bell dynamics for human organs were analysed in [1]. It was suggested that the released energy at resonance was transferred to functions leading to e.g.

- evolution into motorics of the hands
- interaction of speech and brain activity

In the present paper, these concepts will be elaborated in terms of magnitudes for frequencies and occurence in additional applications.

The recovery of sleep is considered extrapolated i.e. translated to actions giving reversed horisontal dumb bells.

The eigen frequencies are based on the results for satellite librations: In horisontal configuration [2], two equal masses oscillates with $3^{\frac{1}{2}}\Omega_p$ where Ω_p is the sidereal rotation of the primary and as derived in [1], while vertical aligned, two masses oscillates with $2^{\frac{1}{2}}\Omega_p$. In the latter case, the masses need not be equal.

2. Change of positions, to rest

An action into rest is to fold hands, Figure 1, since then, the hands are not a horisontal dumb bell, but only one material body.

Figure 1. Folded hands cancel the horisontal dumb bell activity and creates another state.

With Earth as the primary, the periods for the resonance frequencies have the magnitude of diurnals (i.e. 24 hour days), which is large compared with the dynamics of motions for bodyparts. The orientation symmetry is interrupted while lying down, and this may reflect on the brain activity at sleep.

2.1. Sleep, and $\pi/2$ -rotation

$(1-1/2^{\frac{1}{2}})*24=7.0$ hours is the remainder of the diurnal when the time period of a vertical dumb bell is completed. For the horisontal dumb bell, the figure is 10.1 hrs.

The 'action' of sleep is connected to e.g. rest, recovery and dreams, c.f. [3],[4]. Sleeping time is individual and varies around 7-9 hours and during this time, there are two sub-cycles consisting of REM and deep sleep, [4].

When lying down on the back, the pair of eyes is still a horisontal dumb bell if considering the Earth as a primary. Most people fall asleep, when they lie on the side and then, if the head also lies on the side, the eyes become a vertical dumb bell. When closing the eyes, there are no new pictures, but the brain may work with the previous from the day. Then, since reversed dumb bells, the eigen-values are changed and that may result in a different image processing, leading to dreams. The closed eyes may also try to create new pictures with rapid eye movements, REM, using the energy from a vertical dumb bell configuration, and then instead recalling the images from the day and rearranging. Obviously, this results in recovery, but it may also be a creative process when seeing things more random sorted.

2.2. Dramatic resonance

The results are a bit peculiar, compared with usual resonance, when the Earth is the primary. If lying down 7.03 hours, the vertical dumb bell exists exactly a period. Then, the scenario which relates to resonance will affect the primary in the sense that a much larger area covers it, ('a kind of splashing effect'). The 'resonance' for the dumb bell will be that it ceases to exist, and its matter takes another shape. Compared with usual resonance when damping, this is much more dramatic but since the time scales agree, it is probably related.

2.3. Active rest by reconfiguration

An extrapolation of the observations in section 2.1 is that the horisontally aligned parts are at 'active rest' while changing to a vertical alignment.

Although we do not recognise this nowadays, it may be activated by mindfulnes and then beneficial for recovery and useful as a cure.

When the head lies at the side, there is a reconfiguration for the eyes. The hands are easily reversed from a horisontal dumb bell into a vertical, Figure 2.

Figure 2. A reconfiguration of the hands from horisontal into vertical alignment. At left, the limbs are located in a composed horisontal and vertical orientation.

3. Mirrored orientation

A rotation π (half a lap), is found in the TV-series Humle & Dumle, from 1958. The characters consist of the chins with mouths as seen in Figure 3. In fact, the format was accomplished by rotating the picture. Actually lying with the head like that requires activation of several muscles and may be considered an exercise.

Figure 3. The two upside down chins playing 'Humle & Dumle' in a TV-series from 1958.

4. Concluding remarks and discussion

Recalling previous results [2],[1], the dumb bells were quantified with eigenfrequencies.

For example, at lying on the side at rapid eye movement sleep, the eyes and brain may rearrange and process images with another rule, since the eigenvalue has changed.

The symmetry breaking at sleep were considered and similar actions proposed for resting of horisontal dumb bells. To recognise the actions as revitalisation may invoke concentration in terms of awareness and sensibility.

Humans are upright and therefore the brain and the organs obtain the special vertical dumb bell properties, [1]. A mammal with a really short time of sleep is the giraffe, Figure 4 and since very elongated in vertical direction, there are some similarities for the orientations of the organs with a homo erectus, but the sleeping time for the giraffe is only 20 minutes, [4]. It uses the head and neck as a sledge hammer, and to reach food, which are different from the activities of most humans.

Figure 4. Giraffe

An extrapolation of the results in section 2.2, is that sleeping next to a person, figure 5, implies being its dumb bell during the awakening hours, however the real physical configuration do not arrange automatically.

Figure 5. Horisontal dumb-bell consisting of two persons.

Pop Art anti-symmetries

4.

Dumb bell dynamics for collection and redistribution of energy

Summary: Discussion of dynamic models and existing structures, e.g. tree branches, results in designs to manage and limit heavy wind loads. This may rearrange the wind power into other motions, such that the action will not be destructive.

Keywords: dumb bell, Air Balloon, Airship, double pendulum, wind load, design, wind riser, windsock, Koinobori,

1. Review and Introduction

The dumb bells are 3-body problems and the solutions are oscillations for a body of 2 particles. This motion has a kinetic energy. Concerning dancing couples, an overall dumb bell effect may be present, however other dynamics dominate.

For flying objects, it may be notified that an Air Balloon is stable such that it functions in the air layers. A similar non-stable arrangement were found for Airships, when the gondola (passenger-cabin) were too large.

In Chapter 3, the effect was proposed in other contexts e.g. dream activities due to reversed dumb bell orientation and composed configurations, Figure 1.

Figure 1. Brain and body parts in horisontal alignment at rest.

Dumb bell dynamics for human organs were analysed in [2]. It was suggested that released energy at resonance was transferred to functions leading to e.g.

- evolution into motoric of the hands

- interaction of speech and brain activity

- coupling of heart frequency and muscle activities

In the present paper, these concepts will be elaborated for structures, in particular configurations of two pot plants, the double pendulum and devices that reduce wind load. Although well known in mechanics, invoking dumb bell dynamics, e.g. the ratios $3^{\frac{1}{2}}$ and $2^{\frac{1}{2}}$ and possibilities of evolution and adaption, might provide new applications. The main such proposals here concern wind control arrangements.

2. Observations for pot plants

Objects having time constants with the same magnitudes as Earth diurnals are pot plants. For example the exposure to Sun light and need for water supply is repeated. How two plants could be a dumb bell is seen in Figure 2, for a horisontal alignment. These arrangements were common during the 70-ies and the growth were maybe slightly better than for a single plant.

Figure 2. Dumb bell of pot plants, ratio $3^{\frac{1}{2}}$ to Earth diurnal, c.f. [3].

Also a vertical dumb bell of plants, Figure 3, could be expected to energize, however this may result in a polarisation since then, the masses need not be equal. Furthermore, the texture of the plant with roots and crown may be significant and introduce additional d.o.fs that reduces a dumb bell effect.

Figure 3. Vertically aligned dumb bell of pot plants, ratio $2^{1/2}$ to Earth diurnal. The ability to growth may distribute non-uniform, which is an issue for future experimental research in vitro, or in situ field-observations.

3. The double pendulum

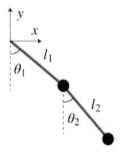

Cohomology with dumb bells: Identification of the upper part as a primary and the lower part as (half of) a vertical dumb bell in its lowest and uppermost position, and in the same way as a horisontal dumb bell at the side positions, the cohomology is established.

The very fast rotations of the lower part can be modelled with nonlinear dynamics and chaos. Invoking dumb bell mechanics with evolution, the resonance gives the fast motion as well as possibilities of growing into a dynamic entity and collect 'free' energy from e.g. bumbs on the ground foundation.

3.1 Interaction with air layers

If the texture of the pendulum has the ability of moving in the wind, it may provide an efficient wind mill. In a windy environment, the chaotic conditions, as well as dumb bell resonance, admits exchange with rotations i.e. angular velocities in adjacent air layers and development of memory. Compared with present wind power plants for electricity production, the velocity is nonuniform and the transmission to generators needs to be redesigned. Since the present arrangements including electric power transmission behave efficient, this will probably not be necessary.

Another application may be to decrease the wind load (or redirect) in stormy weather. If the lower part collects wind energy in layers and increases its rotational velocity, a local estimate of the first law of thermodynamics predicts a decrease in wind energy.

A related problem/subject, not obvious to predict is whether additional d.o.fs in a nonlinear system may provide organisation or additional chaos. For example, the traditional wind mills introduce a harmonic in the wind flows, by its rotation and periodic location, and probably it organises the wind from being extremely high and localised. Generally, for moderate winds, trees act as shields.

Next, we will focus on another application, namely risers to reduce wind loads.

3.2 Wind riser

A pendulum with 2 d.o.f. taken from [4] is shown in Figure 4.

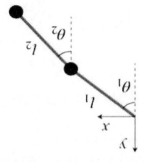

Figure 4. Double pendulum in opposite direction.

Here, we shall consider it a wind riser with the purpose to

- unload the fundament on which it is attached
- decrease the entire wind strength

4. Designs for wind control

Arrangements that may control both moderate and heavy wind are flags of certain shapes. Partly closed cones e.g. wind socks and Koinoboro obviously collect energy into a confined space, c.f. Figure 5.

Figure 5. Wind sock indicating winds.

Concerning design, we may note the new smaller sail-shaped and so-called tear drop flags, c.f. Figure 6.

Figure 6. Tear drop flags that controls the wind.

Likely, a curved side is more endurable than the traditional rectangular.

The shapes confined by the oscillating part of a noncircular orbit, i.e. the eccentricity zone, may be a natural materialisation that oscillates with the wind, c.f. Figure 7.

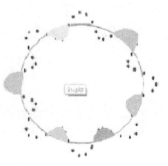

Figure 7. Some colored parts of the eccentricity zone, imagined as flags that may rotate around a stiff circle in the wind.

4.1. Design with cones and bouncing balls

A design with options is cones that rise in the wind. When filled with bouncing balls, e.g. tennis balls, the potential energy, achieved when rising, will redistribute to elastic and damped harmonic bounces. If placed at roofs, the wind loads on the roof shield may be reduced, which could prevent failure.
The same arrangement in a smaller version, with ping-pong balls, may be attached on woven fabric of awnings (sun-shield).

5. Design related to humans (homo erectus)

An idea with designing as dumb bells is that it is consistent with the sub body orientation of an upright person and in the long run, may adjust to work together with homo erectus. In [2], a high hair knot, hat (with pom-pom) and a tie was proposed.

Designs may be accessories introducing dynamics, e.g. hand bags skirts moving when dancing and a tail, c.f. Figure 8 and Figure 9. Some persons, carrying these items, may feel an energy kick by their motions, whereas to other, the opposite downward additional gravity load, is dominating.

Figure 8. Disney fairy-tale with Princess Aurora; Sleeping Beauty, featuring the above mentioned accessories.

Figure 9. Panther with tail.

6. Conclusion

Among other things, dumb bell mechanics were discussed for pot plants and the double pendulum. Pendula were compared with existing designs that move beneficial due to wind energy e. g wind mills and trees. The ultimate shapes of flags to remain and not be thorn in pieces at storms were scrutinised. Finally, some designs to control and reduce wind load were proposed.

Appendix. The governing equation for the free system is given in [4], where also a case for initial acceleration is calculated from the upright non-stable position.

In a windy environment, the pendulum will rise and that can be modelled in several ways. For example, notations from control theory, may be exploited, such that the diagonalised linear system is rewritten (into first order equations), assuming the wind as input.

Theoretically, the inverse response could be analysed and that would be of interest in order to find a way to decrease the wind strength. Since, the above equations are too general and arbitrarily chosen, such a method will not be used.

Instead we will return to the real system:

In a real observation of the angular motion, we may note that the direction of curvature for the riser oscillation is opposite to those of the response when a leaf falls. That could be beneficial if the goal is to reduce the wind, since a change into a hyperbolic system is an option. Such systems have solutions in terms of exponentials, i.e. transients which decrease.

Remark. The branches of a tree moving in the wind, have an opposite curvature, and possibly a reduction and organising effect on the wind.

5.
Noncircular orbits in structural dynamics

Keywords: nonlinear model, second order effects, tangent stiffness, stability, localisation, noncircular orbit, curvature, frequency, aqua plane, Hamiltonian, Glastron speed boat

Introduction

Nonlinear dynamic models give several solutions, among which some provide bounds for the desired or reflect correct behavior. Interactions with boundaries, and quantisation rules from chaos theory and acoustics may be governing. Here we will apply that, in the formats of noncircular orbits to some classical problems in structural dynamics.

Three cases are studied to show the capabilities and limitations.

The first example is a torsional spring. In clockworks, such, together with an oro, which is the moving inertia, gives a steady frequency.

Then, a Finite Element for a Timoshenko beam is developed by constraining the curvature into a noncircular orbit (nco) [1], instead of a circle. This gives a static modification when considering the shape. Intrinsic kinematics will be obtained, keeping the original expression of nco, as a motion.

The third example is shear strain applied as the angular velocity in a noncircular orbit.
This deformation rate will be assumed as a path in a viscoelastic material with localised damage.

I. Torsional spring

II. Beam element

III. Damage at viscoelasticity, for shear rate given by the angular velocity of a nco

I Torsional spring

A torsional spring is a wire with radius r twisted to circles with radius R, in N number of laps. The modulus in the relation between shear stress and angle is $Er4/(8NR)$ where E is the elastic modulus. Consider a harmonic oscillator in a clock-work given by an oro with inertia J, attached to this spring. With the radius of the spring R changed to that of a nco [1], i.e.

$R = r_0 + r_e \sin(fwt)$

The governing equation will be a nonlinear oscillator, and when linearized, the eigen frequency for the harmonic oscillator is $Er^4/(8NR)$ /J with constant R. The correction may be considered as dispersion which drives the device into stable vibrations.

Figure 1. The oro is the wheel below the spring 1, and the regulisator 2. This plane spring differs from that in the model.

In reality, it works such that the oro puts the marker in a new position, and then also gets a kick, since the device released from a wired elastic state. When this occurs at the maximum deflection at one side for the oro, it stays at that angle for (quite) a while, when it push the marker and gets the input kinetic energy.

Hamiltonians for such systems are readily constructed by adding higher order terms, c.f. [3]. From that, the nonlinear input and output Hamiltonian-energies can be identified as integrals of energy over the path, i.e. so-called virials, memory or general higher order terms. The first higher order term in the Hamiltonian with coefficient a_1, c.f. Appendix, is exactly proportional to an integration of potential energy, and the second (with coefficient $-a_2$) could be rewritten with proportionality to a mean of kinetic energy times angle, multiplied with the angle. A more profound analysis of the Hamiltonian (and its dynamics), is an issue for future studies, for example by expanding nco.

II Beam element

For the linear static case, a FE is obtained when entries in the stiffness matrix are multiplied with a factor. For larger angles, the element will be nonlinear. The torsional d.o.f. could be cast in the same format, independently.

In dynamics, there are various options. Since introduced as angles, it is consistent to consider the same for inertia. Another possibility is only for inertia, since nco are a motion. Then, time dependency will give additional inertia and could be formulated as a mass matrix where J is scaled with

$1/(1-2(r_e/r_0)\sin(fwt))$.

Hereby, the output will be internal forced vibrations for the main structure. If also for stiffness, it will be free vibrations which may be damped if not continuously driven by interaction with the main structure.

An application is the shape of pipes from sub water oil wells. Loads from the surrounding fluid could be from wave motions such that parts of the pipe follow the shape and dynamics proposed above rather than the classical beam curvature.

Torsion instability

In axial direction, torsion is considered with an angle applied derived from w in [1].

$\omega(t)=\omega_0 \exp(-2(r_e/r_0)\sin(f\omega_0 t))$...............(1)

For this case, a qualitative model for torsional instability will be derived. Such occurs e.g. at a pre-twinned wire subjected to finite shear strain. (Experiments say that buckling happens when the portion applied is about contraction factor, but this is not verified).

A forward integration of the angular velocity (1) gives

$q= q_0 +Dq + r_e/r_0(1/f)\cos(fDq)$ where Dq is an incremental angle.

Consider a quasi static case described with the potential $V(q-q_0) =V(Dq)$ where V is a quadratic function of Dq.

A new angle is applied incrementally, such that $V(1)=V(0) + V'dq$

where V' = 2Dq(0)(1-(re/r$_0$)sin(fDq(0))) with a forward evaluation and dq is the differential (or subincrement) of Dq. Stability is lost when the tangent 1/V' is inf-large i.e V'=0, which gives Dq(0)=(r$_e$/r$_0$)1/f for small Dq(0).

Remark. Depending on how time is chosen, the trigonometric term could be different.

This simplified constrained approach gives a qualitative description. Since twinned, the concept is 3-dimensional and there are nonlinear interactions between the deformation paths before the instability. The eigen-value for a stress state of pure shear is at pi/4 from the axial direction. A detailed explanation of the bifurcation could be a mode shape of pressure and normal strains along that curve circumferential. Then, normal stress will 'explode' it at the bifurcation. In terms of Euler buckling, the angle depends on the winding and upper scale pressure in beam direction which is imposed at additional twinning, in some way which could be quantified with detailed modeling. Or this result is the truth, such that chaos divides into hypersurfaces which are independent, and the interaction from other dimensions are limited and quantised, mostly determined by intrinsic properties in the current dimension. The interaction may be given as a suitable quantised amount of pressure impact or energy, which organizes in the dofs of the receiving space independently.

III Damage at viscoelasticity, for shear rate given by the angular velocity of a noncircular orbit

In [2], fatigue is modelled with damage when a stress measure is above zero a limit value. The format is general and invokes viscoelasticity, a nonlocal formulation of the yield criterion and FE implementation.

Here, this model will be evaluated for a certain deformation, and exact results are provided.

Preliminaries. Shear rate is assumed as w, for a noncircular orbit (1). Such deformation paths may be present at subsurface cracking due to wear, and at soil slope stability.

Cumulative damage at fatigue loads of subsurface cracks due to wear.

Linearising and assuming $k_\sigma=-2$, the damage law (15) in [2] can be integrated to read $D(t)=(D_0-1)\exp(-F(\sigma)Dt)+1$ where $D(t)$ is damage, Dt is the time step and $F(\sigma)$ is a constant linearized function of stress and D_0 is the constant damage (cumulated) from previous. In [2], $k_\sigma=0.02$ is used for ice, and k_σ around 1 and 2, to model two structural materials.

Growth of damage is active when the state is above a limit, and depends on the stress and Dt which from the constitutive equation (1) in [2] depends on the strain rate and shear modulus with viscoelastic parameters and time integration (convolution).

A bound for Dt is determined from time above the damage limit. Assuming that the mean value of w ; w_0 is below, gives

$Dt<\pi/(w_0f)$ ....(2)

Hereby, maximum growth is given by (1), with Dt as the upper limit of (2).

Remark. It could be expected, for some rate dependent material, that the growth would increase with the frequency w_0, if the material does not have time to relax in between. Revisiting (1) in [2], if relaxed, stress is zero and this will decrease Dt. Hereby the bound (2) will be decreased to $Dt<\min(pi/(w_0f), \lambda)$ where λ is the effective relaxation time in the shear modulus. Therefore, present viscoelasticity will decrease the damage effect.

In details:

The constant term do not contribute to damage if below the limit

i) Fast relaxation time, i.e. large time constants in the shear modulus and $G_{inf}=0$, will give the initial value of shear rate, but scaled. This has the upper bound $(2r_e/r_0)w_0$ and hereby the factor f do not contribute additional to fatigue, and the stress will be relaxed rapidly, determined by the effective viscous time constant and the above value giving decreased damage.

ii) Medium relaxation. Stress consists of the integration of a decreasing harmonic, which gives $\exp(-t/\lambda_i)$ (decreasing harmonic), where the harmonic depends on both w_0,f, and viscoelasticity. This will be further analysed in the forthcoming subsection.

iii) Slow relaxation. When shear modulus is constant i.e. $G=G_{inf}$, the integration will give the same harmonic with a phase. Then, when not linearising the damage law, damage will achieve the same format as (1), which gives damage growth as a step, and a superimposed harmonic while positive i.e. for a time $pi/(w_0f)$. This is the case with maximum damage.

Exercise. Evaluate the integrals in [2], to obtain the exact expressions for stress and damage function D(t).

Subscale dynamics

In [2], consistent linearisation is used while integrating the rate equations. Here we will keep the nonlinear dependency, in order to evaluate the dynamics during damage transition.

Another linearization of (15) gives $D(t) =C\sigma(t)$, i.e. damage rate directly proportional to stress. Hereby D(t) will oscillate, which is not a spurious result, but rather to be interpreted as instability, nonlinearity and additional information of the fine structure of the model.

The new 'minor scale time constant' depends on both $1/(w_0f)$ and viscoelastic decrease time constants, λ_i. Also resonance with some oscillations in the structure could be a possible scenario. The entire phenomenon is that of intrinsic dynamics on a minor scale, which may promote fracture.

Extrapolation. Assuming this subscale dynamics as the point of departure, such could induce a shear rate process on a superlevel. This may be interpreted as damage, which induces shear at an above level and so on. Such an extrapolation upwards will give higher dimensions since the first damage was scalar and then shear occupies two spatial dimensions.

In conclusion. Shear stress at localisation into a noncircular orbit for a viscoelastic material was evaluated. A boundary value problem was not solved, but instead a crack path consistent with only shear rate was assumed. At transition for cumulative damage, a superimposed oscillation was proposed, during the main transition times, which implies a fine scale.

(Assuming that such a scalar variation induces the 2-dimensional strain state on another level which also oscillates (as forced vibrations if coupling between

43

levels), and in turn acts as 2 sublevels to new superleves and so on, gives a spatial multidimensional space of shear cracks.) Returning to the initial spatial dimensions with a scalar damage, the model describes texture formation in terms of decreased stiffness in layers of a visco-elastic material behind a rotating sphere or planet.

Other recent models intended for shear slip are NURBS at elasto-plastic inclusions [4].

Acknowledgements. To Dr Marco Petrangeli and Mr Kondrashov for valuable discussion during the course of this work.

References

Chapter 1

[1] Strömberg L (2015). Motions for systems and structures in space, described by a set denoted Avd.Theorems for local implosion; Li, dl and angular velocities. Journal of Physics and Astronomy Research, 2(3): 070-073.
[2] Strömberg L and Soualmia A (2018). Sensing Sound as a Temperature with Integrative Hair Skin Physiology, Technical report Researchgate
[3] https://www.researchgate.net/project/Quantum-Field-Theory-What-are-virtual-particles/update/5b57475cb53d2f89289b8a09

[4] Strömberg L and Soualmia A (2018). Helix trajectories in dynamics of nco, duality & materialization. Technical report Researchgate

[5] Goldstein, Classical Mechanics

Chapter 2

[1] Strömberg L (2018).The Matlab-and Octave-matrix magic(n). Analysis and applications.
[2] Strömberg L (2017). Celestial geometry and evolution of Homo Erectus .
[3] Strömberg L (2018). Sub-bodies and space times for homo erectus considered as fractal systems. DOI: 10.13140/RG.2.2.13348.58248/1

Chapter 3

[1] Strömberg L (2017). Celestial geometry and evolution of Homo Erectus

[2] Klemperer W.B and Baker R.M. (1957). Satellite Librations, Astronautica Acta, 3: 16-27, Springer.

[3] https://sv.wikipedia.org/wiki/S%C3%B6mn

[4] Everdahl G (2005). Bildningsakuten, Bonnier Carlsen, pp.76-77.

Chapter 4

[1] Strömberg L (2016). Genus perspective of a female homo erectus, visualized by a torus and compared with vertical dumb bells in tidal resonance. Journal of Scientific Research and Essays 1(2).
https://www.pearlresearchjournals.org/journals/jsre/archive/2016/Feb/Abstract/Stromberg.html

[2] Strömberg L (2017). Celestial geometry and evolution of Homo Erectus

[3] Klemperer W.B and Baker R.M. (1957). Satellite Librations, Astronautica Acta, 3: 16-27, Springer.

[4] Gerck E.V. and Gercky E. (2019). From Newton to the Lagrangian and Hamiltonian Formalism: Euler-Lagrange equation, Noether Current, and Conservation Laws. Planalto Research

Chapter 5

[1] Strömberg L (2015). Models for locations and motions in the solar system. Journal of Physics and Astronomy Research, 2(2): 062-066

[2] An equivalent stress-gradient regularization model for coupled damage-viscoelasticity. Article in Computer Methods in Applied Mechanics and Engineering · May 201 DOI: 10.1016/j.cma.2017.04.010 Juan G. Londono, Luc Berger-Vergiat, Haim Waisman

[3] Lena J-T Strömberg, RATIONAL POINTS ON ELLIPTIC CURVES; THE ULTIMATE SOLUTION OF THE MILLENIUM PRIZE PROBLEM; BSD-CONJECTURE, *SCIREA Journal of Mathematics*. Vol. 1, No. 1, 2016 , pp. 175 - 183 .

[4] Gernot Beer, Vincenzo Mallardo, Eugenio Ruocco, Benjamin Marussig, Jurgen Zechner, Christian Dunser, Thomas-Peter Fries (2017). Isogeometric Boundary Element analysis with elasto-plastic inclusions. Part 2: 3-D problems. Comput. Methods Appl. Mech. Engrg. 315 (2017) 418–433

Crochet description
23 lm, fold to a ring with 1 sm.
Lap 1: 10 lm, 1 sm in the 6th m of the ring, 5lm, skip 10 lm, 1 sm in the next, 10 lm, skip 5 lm, 1 sm in the ring (i.e. the beginning of the lap).

This gives the pattern in the figure above, which is similar to a nco with f=3/2, c.f. Figure 1 in Chapter 1. For larger integer f, a shape is even more easily accomplished, e.g. as a symmetric pattern around a ring, c.f. Figure 7 in Chapter 4.

'Once upon a dream': Princess Aurora, from Figure 8, Chapter 4, now with her Prince. A realisation of a dumb-bell configuration adressed in the end of Chapter 3.